CUBE GAMES

CUBE GAMES

92 PUZZLES & SOLUTIONS

Don Taylor and Leanne Rylands

An Owl Book

Holt, Rinehart and Winston
New York

First published in the United States in 1981 by
Holt, Rinehart and Winston, 383 Madison Avenue,
New York, New York 10017.
Published simultaneously in Canada by Holt, Rinehart
and Winston of Canada, Limited.
Library of Congress Catalog Card Number: 81-84327
ISBN: 0-03-061524-0

First American Edition
Designer: Andrena Millen

Printed in the United States of America
10 9 8 7 6 5 4 3 2 1

Plus ça change, plus c'est la même chose.

Contents

CUBE
GAMES

Introduction

In 1980 it was still possible to believe that
Rubik's Cube was just the latest puzzle craze —
like the yo-yo, destined to flower briefly then fade
away. But one year later the demand for the cube
continues unabated. It is no longer sensible to
think of it as a short lived wonder. There must be
something more to it.

At first glance it appears to be a disarmingly
innocent toy. It is only after twenty minutes or so
of diligent twisting that one realizes how
frustratingly difficult it is to keep track of the way
the patterns are changing. Some people give up at
this point and stop trying to get the cube back to
its original state. But most of those who pick up
the puzzle become addicted to searching for the
solution, not knowing how exceedingly difficult
this can be. Fortunately there are solutions
available in books such as **Mastering Rubik's
Cube**.

Being able to pick up a scrambled cube and
restore it to its original pattern is certainly very
satisfying. But it is only the first stage of the
quest to discover the mysteries of the cube.
Making pretty patterns is the next stage. That is
what this book is all about. Even if you haven't
read **Mastering Rubik's Cube** you will still be
able to follow our instructions for making the
patterns. But you will probably find it a lot more
enjoyable if you know how to unscramble a
messed-up cube, since this makes it easier to fix
up any errors you might make. (One way of doing
this is described in Chapter 10.)

The mysteries of the cube certainly don't end
with the patterns. Quite early in its career,
Rubik's Cube attracted the attention of
mathematicians. Many articles have been written
linking it with the 'theory of groups'. It is now
regarded as one of the most fascinating teaching
aids ever invented for illustrating the mathematics
of symmetry.

Apart from being a teaching aid, the cube is an extremely colourful ornament, even a work of art. In fact, the New York Museum of Modern Art has already included it in its permanent collection. No doubt it is this beguiling combination of aesthetic and intellectual appeal that explains its attraction, despite the extraordinary difficulty of finding a solution.

In 1974, at the age of 29, Ernö Rubik, a Budapest architectural engineer, made his first cube. The cube was made of wood. In 1977, the Hungarian Polytechnika Cooperative began producing a plastic version. In 1978 it won a prize at the Budapest International Fair, and cube mania began in earnest. The Ideal Toy Company obtained world distribution rights, and addiction to the puzzle spread across Europe to Britain, then over the Atlantic to the United States. Within a few years tens of millions of cubes had been sold.

Other manufacturers have modified the design (but not the internal mechanism) to produce 'octagonal barrels', 'cuboctahedra' and other strange shapes. There are many sizes of cubes available, including tiny cube pendants and various models that can be attached to a key ring. All of the pattern instructions in this book can be applied to these new models. The details on how to cope with these variations can be found in Chapter 11.

There are many ways that a manufacturer can colour a cube. Even if only plain colour patches are used, there are still thirty ways of arranging them on the cube. We have chosen a particular arrangement for our illustrations. It is quite possible that this won't correspond to the one on your cube. You needn't worry about this because the *instructions* don't change. All that will happen is that you will produce the same patterns with a different colour arrangement.

Only three faces of a cube can be seen at the one time. So the illustrations don't show you the whole pattern; you have to guess the arrangement of the other faces. Most of the time it is quite clear what the rest of the pattern should be. But the best way to appreciate the patterns fully is to carry out the instructions on a cube of your own.

We would like to think that for most of the patterns we have found the shortest way of getting there. But that is probably only wishful thinking on our part, so this book presents two challenges to the enthusiastic reader. The first is to find new, interesting patterns. The second is to improve upon our methods and find better ways of getting them.

We cannot close this introduction without recording our gratitude to David Singmaster. His book **Notes on Rubik's 'Magic Cube'** (Enslow Publishers, Hillside N.J. 1981) has been an inspiration and a great help to us. We heartily recommend it to anyone who wishes to go more deeply into the magic of the cube.

1. The Language of Patterns

If you have already looked at a few of the illustrations in this book you may have noticed that the instructions for making the patterns are given in a kind of shorthand. This chapter tells you how to read that shorthand. But before you can make use of it you will need to be able to get your cube back to its original state in which all nine squares on each face are the same colour. This is the *plain pattern* shown in Figure 1 and it is the starting position for most of the patterns that we describe later on.

We shall get on with the shorthand in a moment, but first of all you need to know how to unscramble a chaotic cube. One way of doing this is to take the whole thing apart and reassemble it so that all the small cubes are back in their correct positions. The instructions for this method can be found in Chapter 10, towards the end of the book. However, most people feel that doing it this way is cheating and that it is not the sort of thing you should do whilst other people are watching. A more satisfying way of restoring the cube is to find a solution of your own — or to follow the instructions in **Mastering Rubik's Cube**.

The book **Mastering Rubik's Cube** also contains a simple way of describing sequences of moves of the cube. We have used exactly the same method to describe the patterns in this book. It is this shorthand that we use throughout the book. In order to read our descriptions of the puzzles and our instructions for making the patterns you will need to know what it all means.

The basic idea is to use a letter of the alphabet to represent quarter turns of the faces and then to describe compound moves by 'strings' of letters.

Figure 1

Place the cube in front of you and think of the six faces as *up, down, right, left, front* and *back.* (See Figure 2.) This choice of words to describe the faces was introduced by David Singmaster in his book **Notes on Rubik's Magic Cube** (Enslow Publishers, Hillside N.J. 1981). It is also used by Douglas R. Hofstadter in his article on the cube which appeared in the March 1981 issue of the magazine **Scientific American.** The initial letters of these words will be used to indicate a *clockwise quarter turn* of the corresponding face. In order to understand what *clockwise* means imagine each face to have a clock at its centre; then a face is said to move clockwise when it moves in the direction of the hands of its clock; the opposite direction is known as *anticlockwise*. It is worthwhile thinking about this fairly carefully and referring to Figure 3 whenever you are in doubt, since many people find that the direction of a clockwise turn of the back, left or down faces, according to the description just given, is in conflict with their intuition.

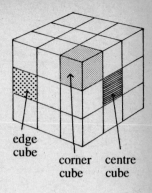

edge cube

corner cube

centre cube

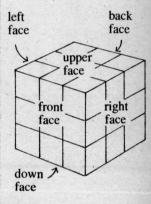

left face

back face

upper face

front face

right face

down face

Figure 2

5

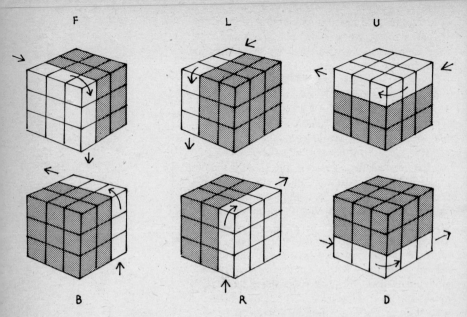

F   L   U

B   R   D

Figure 3

To resume: clockwise quarter turns of the *up, down, right, left, front* and *back* faces are represented by the letters U, D, R, L, F and B. To indicate an *anticlockwise* quarter turn put -1 after the letter. To indicate a *half-turn* (that is, *two* quarter turns) we put 2 after the letter. A *sequence* (or *string*) of moves is indicated by writing the letters one after another. For example, the string of letters BL^2R^{-1} means that you should rotate the *back* face through a clockwise quarter turn, then rotate the *left* face through a half turn and then rotate the *right* face through an anticlockwise quarter turn.

Notice that the faces of the cube you are calling *up, down, right, left* and so on, depend on how you *hold* the cube. So while you are carrying out a sequence of moves you should take care not to change the *orientation* of the cube before you have finished the sequence. For example, the face which begins at the front should remain at the front throughout the sequence. This hazard could be avoided by using the colours of the centre cubes as the names of the faces. But not all cubes are coloured alike, so it would not be a good system to use in this book. The great advantage of the naming system that we have chosen is that the description of the moves remains the same, no matter how your cube is coloured. It also applies to the various modifications of the cube such as the octagonal barrel (Figure 91).

Suppose that you tried out the example BL^2R^{-1} that we gave above, but now you would like to undo the effect of this and get your cube back to the state it was in before you applied the move. All you have to do is to use the move RL^2B^{-1}. That is, to carry out the *reverse* of a string of moves you simply begin at the right hand end of the string and reverse each separate move, remembering that the reverse of a half turn is still a half turn since it does not matter whether it is performed clockwise or anticlockwise. We already use -1 after a *letter* to indicate the reverse effect of that letter and in the same way we shall use -1 after a whole *string* to indicate the reverse of the string. So the reverse of BL^2R^{-1} is $(BL^2R^{-1})^{-1}$. As we said before, this is just RL^2B^{-1} (because the reverse of R^{-1} is R, the reverse of L^2 is L^2 and the reverse of B is B^{-1}).

We can also use a number after a string to indicate *repeats* of the string. For example, $(BL^2R^{-1})^3$ means that BL^2R^{-1} is to be repeated *three* times: it is the same as performing $BL^2R^{-1}BL^2R^{-1}BL^2R^{-1}$.

This can be a useful aid if you wish to memorize long strings of moves. A pattern of two plain faces and four faces with crosses (Figure 4) can be produced by a string of twenty-four moves. Even though this is rather long, it is easy to remember: it is just FBRL repeated three times, followed by LRBF repeated three times. We can write this as $(FBRL)^3(LRBF)^3$.

But a word of caution before you attempt to carry this out. It is a rather long string of moves. If you have not had much practice in doing this sort of thing, it is depressingly easy to lose track of where you are and to end up with a thoroughly scrambled cube. Instead of subjecting yourself to the possibility of a chaotic cube you could turn to the next chapter and improve your skills by practising patterns taking no more than eight

$(FBRL)^3(LRBF)^3$

Figure 4

moves. But for those brave souls who cannot pass up a challenge, here is something else to try.

Beginning with the cube in its original state (six plain faces) apply the seven half turns $L^2R^2D^2L^2R^2F^2B^2$ *and then find* **three** *more half turns which take the cube to the 'four cross' pattern shown in Figure 4.*

The solution to this problem can be found in Chapter 5.

2. Simple Beginnings

To get you started, here are some appealing patterns which are quite simple to make and easy to remember. One of the simplest of them all is shown in Figure 5. It is made up of six 'diagonal crosses', one for each face. It can be obtained by starting with the cube in its original state (with six plain faces) and applying the six half turns $U^2D^2R^2L^2F^2B^2$ (or variations such as $R^2L^2U^2D^2B^2F^2$).

From this pattern it is rather easy to get back to the plain pattern: you just apply the same string of moves $U^2D^2R^2L^2F^2B^2$ and there you are. This is something of an exception though, and usually you need to *reverse* the string (as described in the last chapter). In this case the reverse is $B^2F^2L^2R^2D^2U^2$ and of course this also takes you from the pattern of Figure 5 to the plain pattern.

Another pretty pattern which is easy to make has six 'spot' faces (Figure 6). One way of getting it is to apply the string $FB^{-1}UD^{-1}RL^{-1}FB^{-1}$ to the plain pattern of Figure 1.

This move takes the ring of eight squares surrounding the central square of the front face to the upper face, the eight squares (except the central one) of the upper face to the right face and the corresponding eight squares of the right face back to the front face. And the same sort of thing happens to the down, left and back faces. To get back to the original state apply the *reverse* move $(FB^{-1}UD^{-1}RL^{-1}FB^{-1})^{-1}$. Remember that to reverse a string of moves you begin at the right hand end and reverse each move in turn. In this case we get the string $BF^{-1}LR^{-1}DU^{-1}BF^{-1}$ and sure enough this takes us back to where we began.

Notice that FB^{-1} (as well as RL^{-1} and UD^{-1}) moves opposite faces in such a way that the colours on the four corner squares of each face remain the same. This is a *slice* move and more will be said about it in the next chapter.

An easy *stripe* pattern which can be made with

$U^2D^2R^2L^2F^2B^2$
Figure 5

$FB^{-1}UD^{-1}RL^{-1}FB^{-1}$
Figure 6

9

$(F^2R^2B^2)^2$

Figure 7

just six half turns is shown in Figure 7. The
formula for it is $(F^2R^2B^2)^2$. Remember that the 2
at the end of the string means that $F^2R^2B^2$ is to be
performed *twice*. Can you work out how to get
back to the plain pattern? This is just the problem
of finding the *reverse* of the given string of
moves.

We can now give a rather straightforward
puzzle which also involves this pattern. All of our
puzzles will be numbered from now on and the
solutions to them will be found at the back of the
book.

1. *Using just* two *half turns can you change the
stripe pattern of Figure 7 into the stripe
pattern of Figure 8?*

This pattern and others like it are described in
Chapter 6.

Yet another easy pattern is shown in Figure 9.
As you can see it is a simple combination of spots
and stripes. To make it all you need to do is to
begin with the plain pattern and apply $B^2D^2U^2F^2$.

Figure 8

$B^2D^2U^2F^2$

Figure 9

10

Another way of arriving at essentially the same thing is to use the string $U^2F^2B^2D^2$. If this string is combined with the previous one we get $B^2D^2U^2F^2U^2F^2B^2D^2$. When this is applied to the plain pattern, we come upon the pattern of Figure 7, this time from a different angle.

Here is another puzzle to test your skill and to whet your appetite for things to come.

2. *Begin with the plain pattern and apply the string $U^2D^2L^2U^2D^2R^2$ to create the pattern shown in Figure 10. Now see if you can find a way of using the pattern of Figure 9 to change this into the* 'four spot' *pattern shown in Figure 11.*

$U^2D^2L^2U^2D^2R^2$

Figure 10

So far we have seen a six spot pattern and a four spot pattern. What about a *two spot* pattern with spots on opposite faces such as the one in Figure 12?

This could be a tantalising question if left unanswered. But this chapter is meant to be fairly simple. So, rather than maintain the suspense a moment longer than necessary, we prefer to tell you that it is *impossible* to make such a pattern. And quite a few of the puzzle patterns in other chapters will turn out to be impossible too! In the case of the 'two spot' pattern there is a fairly simple reason why it cannot be made, so we conclude this chapter with a question.

3. *Why is the two spot pattern impossible?*

Figure 11

Figure 12

3. Slice Patterns

FB⁻¹

Figure 13

In the last chapter all of the patterns were created using either half turns or the *slice moves* FB⁻¹, RL⁻¹ and UD⁻¹. Now we shall turn our attention to what can be achieved using *only* slice moves. The number of patterns you can make using these moves is not too large. If you scramble the cube with slice moves it can never get very far from the plain pattern. This is because a slice move really only changes the cubes of a 'middle slice' as shown in Figure 13.

The corner cubes never actually get scrambled and the edge cubes remain in their 'slices'.

Even at its most scrambled (by slice moves only), the pattern on each face is still very symmetrical. The colours on the four corner squares of a face must all be the same and the colours on opposite edge squares also match (as in Figure 14).

Figure 14

There are twenty-four different ways that you can choose to hold your cube. (This is because the cube has *six* faces, and any one of these can be chosen to be the *upper* face. Once this has been done, there are still *four* choices left for the face which you hold to the front.) You might think this would mean that for each possible pattern you could get twenty-four variations by changing the starting position of the cube. It turns out that this does not always happen. For some strings of moves, such as $U^2D^2R^2L^2F^2B^2$, which produces the pattern of Figure 5, you set the *same* pattern no matter how you start. (Notice that U^2D^2 has the same effect as $(UD^{-1})^2$ so it really is a slice move.) As an example of the sort of variations which can occur, have a look at the two 'six spot' patterns in Figure 15.

Both these patterns can be made by applying the slice moves $FB^{-1}UD^{-1}RL^{-1}FB^{-1}$ to the cube. The only difference between them is that for the first one, we began with orange on the upper face and yellow on the front face, whereas for the second, we had red on the upper face and yellow on the front, and we turned the cube upside down after we applied the string of moves. So these patterns should be thought of as essentially the same. The number of *essentially different* patterns you get, using slice moves only, is *fifty*. Most of these are uninteresting, and they include such things as the pattern of Figure 13 (created by FB^{-1}) and the pattern of Figure 16 (created by $(RL^{-1}F^{-1}B)^2$).

Figure 15

$(RL^{-1}F^{-1}B)^2$

Figure 16

R²L²UD⁻¹F²B²UD⁻¹
Figure 17

To check that there *are* only fifty patterns is exceedingly tedious and we certainly do not intend to go into that here.

A few of the more elegant slice patterns already occur in the previous chapter. For the 'six crosses' pattern see Figure 5 and for the 'six spot' pattern see either Figure 6 or Figure 15. The 'four spot' patterns of Figure 11 was created entirely by half turns. However it is also possible to get it using only slice moves. A string of moves to do this is R²L²UD⁻¹F²B²UD⁻¹ and the result is shown in Figure 17.

The 'four spot' pattern followed by the 'six crosses' produces the pattern of Figure 18. Another way of getting this is to use F²B²U⁻¹DF²B²UD⁻¹.

By applying the 'six spot' string to the pattern of Figure 18, you arrive at the pattern of Figure 19. Alternatively, this comes from the 'six spot' pattern combined with the 'six crosses' pattern and also from the string FB⁻¹UD⁻¹R⁻¹LU²D²F⁻¹B.

And now for a puzzle.

F²B²U⁻¹DF²B²UD⁻¹
Figure 18

4. *Begin with a 'six spot' pattern, turn the cube around in your hands a few times and again apply the string which creates the 'six spot' pattern. Only three types of pattern can emerge at the end of this process. What are they?*

RL⁻¹UD⁻¹FB⁻¹U²D²R⁻¹L
Figure 19

14

It is always fairly easy to get back from any slice pattern to the plain pattern. All you have to do is to head for a 'six spot' or 'four spot' pattern and then use one of the processes you have learnt in this chapter to unscramble the spot pattern. But how do you get to the spot pattern? The idea is to regard the corner cubes as fixed and move the middle slice between the right and left faces, say, so that the edge cubes at top of the front and back faces match the corner cubes of the upper face. Then the upper face will at least have the squares along the front and back edges matching. And if you are lucky, you may even have a spot or plain face as shown in Figure 20.

If you have not yet produced a spot or plain pattern on the upper face, move the slice between the front and back faces until you do so. To finish off, move the slice between the up and down faces until you create spot or plain patterns on the remaining faces. Then unscramble the spot pattern. This can make a good party trick. If you are careful to use only slice moves to scramble the cube, you can use the method just given to 'solve the cube' and impress your friends with your speed!

Once you have mastered solving the cube using only slice moves you should have no trouble with the next few puzzles.

Figure 20

15

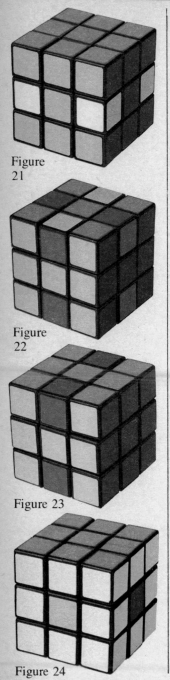

Figure
21

Figure
22

Figure 23

Figure 24

5. *Is it possible to make the pattern shown in Figure 21 in which the edge cubes of the middle slice between the up and down faces are each moved one position around the slice and all the other cubes remain in place?*

6. *Using only slice moves, it is possible to make the pattern of Figure 22 in which the edge cubes of the slices between the front and back faces and between the right and left faces are moved around one position leaving all the other cubes in place?*

Surprisingly, you may find Puzzles 6 and 7 much easier than Puzzle 5, even though at first glance they appear more complicated, and look as though they could be solved by finding a string of moves for puzzle 5 and repeating it.

7. *Can you find a string of slice moves that creates the pattern of Figure 23 in which each edge cube between the right and left faces is swapped with the diagonally opposite one?*

And finally, here is an impossible pattern for you to contemplate. It is a 'four spot' pattern, but it is not quite the same as the one we made before (Figure 17). Can you see the difference?

4. Corner Cube Manoeuvres

Instead of continuing our catalogue of patterns, we shall fill the next few pages with illustrations of some short strings of moves which affect only corner cubes. All the moves described in this chapter will be '3-cycles'. This means that the effect of each string is to 'cycle' just *three* corner cubes around and leave all the other small cubes in place. Apart from improving your ability to unscramble the cube, the knowledge of these moves should enable you to create interesting patterns of your own.

For each '3-cycle' of corner cubes we have done our best to find the shortest possible string of moves that has the desired effect. We do not claim that these moves cannot be improved upon, so we ask the obvious question:

Can you find a shorter way of doing any of the corner cube moves described in this chapter?

We begin with eighteen strings of moves which affect only the corner cubes of the upper layer. These strings are grouped into nine pairs in such a way that each move is the *reverse* of the other move of its pair. The diagrams show the effects that these strings of moves have on the *plain pattern*. If you want to get *back* to the plain pattern starting from one of the positions shown here, choose the pair of patterns which includes your pattern and apply the *reverse* string; that will be the string under the *other* pattern of the pair.

$R^2DR^{-1}U^2RD^{-1}R^{-1}U^2R^{-1}$

$RU^2RDR^{-1}U^2RD^{-1}R^2$

Figure 25

17

$R^{-1}ULU^{-1}RUL^{-1}U^{-1}$ $ULU^{-1}R^{-1}UL^{-1}U^{-1}R$

Figure 26

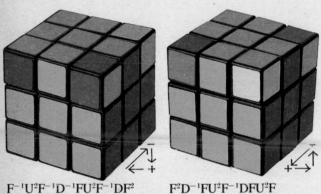

$F^{-1}U^2F^{-1}D^{-1}FU^2F^{-1}DF^2$ $F^2D^{-1}FU^2F^{-1}DFU^2F$

Figure 27

$RBR^{-1}FRB^{-1}R^{-1}F^{-1}$ $FRBR^{-1}F^{-1}RB^{-1}R^{-1}$

Figure 28

FD²B²DF²D⁻¹B²DF²DF⁻¹

FD⁻¹F²D⁻¹B²DF²
D⁻¹B²D²F⁻¹

Figure 29

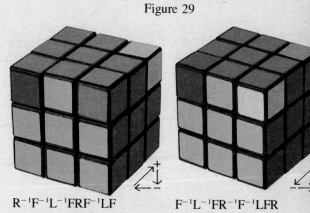

R⁻¹F⁻¹L⁻¹FRF⁻¹LF

F⁻¹L⁻¹FR⁻¹F⁻¹LFR

Figure 30

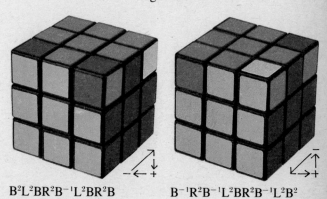

B²L²BR²B⁻¹L²BR²B

B⁻¹R²B⁻¹L²BR²B⁻¹L²B²

Figure 31

19

LF²LB²L⁻¹F²LB²L²   L²B²L⁻¹F²LB²L⁻¹F²L⁻¹

Figure 32

F⁻¹LF⁻¹R²FL⁻¹F⁻¹R²F²   F²R²FLF⁻¹R²FL⁻¹F

Figure 33

20

As well as changing the positions of the corner cubes these moves also *twist* the small cubes in various ways. This information is contained in the small diagrams made up of arrows and + and − signs which accompany the illustrations of the patterns. The + means that the corner cube can be thought of as being twisted 'clockwise' after it has been moved into position (see Figure 34). Similarly, the − sign indicates an 'anticlockwise' twist of the corner cube after it has been moved into position.

The remaining figures in this chapter present a selection of 'corner 3-cycles' which affect cubes in more than one layer. Not all possibilities of twists have been shown here, but more information about 'twisting and twirling' is given in Chapter 8.

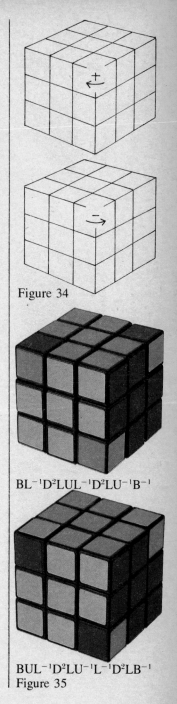

Figure 34

BL⁻¹D²LUL⁻¹D²LU⁻¹B⁻¹

BUL⁻¹D²LU⁻¹L⁻¹D²LB⁻¹

Figure 35

21

$RBL^2B^{-1}R^{-1}BL^2B^{-1}$

$BL^2B^{-1}RBL^2B^{-1}R^{-1}$
Figure 36

$BL^{-1}RBR^{-1}FRB^{-1}$
$R^{-1}F^{-1}LB^{-1}$
Figure 37

Even though we haven't shown all possible 3-cycles of corner cubes it turns out that there is a neat way of constructing them from the moves given in Figures 25 to 33. Suppose you want to 'cycle' three corner cubes but the corners you are interested in are not all in the same face. First bring the corner cubes you want to 'cycle' into the three positions of the upper face that we already know how to cycle (the front positions and the back-right position). Do this any way you please and call the move you use to do it X. Of course by applying X you will rearrange lots of other cubes, but don't worry about this, they will be fixed up later. (So that you *can* fix things up it is wise to write down the move X. Later on you will need to apply the *reverse* of X.)

Now use one of the upper face 3-cycles, for example $RBR^{-1}FRB^{-1}R^{-1}F^{-1}$ of Figure 28 and call this move Y. To finish off, apply the *reverse* of X. This will put all the cubes back to their correct positions, except that the three corner cubes you selected will now be 'cycled' around. What you have done is to apply XYX^{-1}. This move is called a *conjugate* of Y. It transfers the effect of Y to a new collection of corner cubes and as you can see this can be quite useful.

As an example, suppose X is BL^{-1}, then X^{-1} is LB^{-1} and if Y is the same as before, then XYX^{-1} is $BL^{-1}RBR^{-1}FRB^{-1}R^{-1}F^{-1}LB^{-1}$. If you apply this you will find that it cycles the three corner cubes at the positions shown in Figure 37.

5. Crosses

There are two sorts of cross patterns that can appear on a face of the cube. One is the 'diagonal' cross in which the centre and corner squares form an X, as shown in Figure 38(i). The other is the 'straight' cross, or 'plus' pattern in which the centre and edge squares form a +, as shown in Figure 38(ii).

It is possible to produce patterns in which the two sorts of crosses appear together. An example of this is the pattern of four + faces and two X faces shown in Figure 39.

8. *Can you find a shorter way of producing the pattern in Figure 39?*

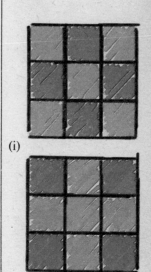

(i)

(ii)

Figure 38

$L^2R^2B^2U^2D^2L^2U^2D^2R^2F^2$

Figure 39

23

As well as combining the two sorts of cross patterns we can combine crosses with other types of patterns. A pattern with two + faces, two spot faces and two plain faces is shown in Figure 40.

Examples of crosses combined with stripes can be found in the next chapter. For the rest of this chaper we concentrate on patterns which have only cross or plain faces.

9. *We have not bothered to give the instructions for producing a pattern with four plain faces and two + faces. Is this because the move is very easy or is it perhaps impossible?*

$F^2B^2D^2F^2B^2U^2L^2U^2D^2R^2$
Figure 40

In any pattern of four + faces and two plain faces, the plain faces must be opposite one another. One way of making a pattern like this was described towards the end of the first chapter. The question as to how this could be done using exactly ten half turns was left unanswered there. A solution to this puzzle is presented here in Figure 41.

10. *The pattern in Figure 42 has two plain faces and four + faces yet it is not quite the same as the pattern in Figure 41. Can you see the difference and can you find a way of creating the pattern of Figure 42? (Hint. Try combining the string $(R^2F^2B^2)^2$ of Figure 50 with U and U^{-1}.)*

$L^2R^2D^2L^2R^2F^2B^2U^2F^2B^2$
Figure 41

There are two types of patterns with six + faces. These are shown in Figures 43 and 44. In Hofstadter's **Scientific American** article (March 1981) these were called the 'Christman cross' and the 'Plummer cross'.

Figure 42

If you combine a 'Plummer cross' with a 'Christman cross' you will get a solution to Puzzle 10, but this is a rather lengthy way of solving the puzzle.

All of the remaining patterns in this chapter have two, four or six X faces and the other faces plain. As you can see, one way of making a pattern with six X faces is very easy and occurred before as Figure 5. The other six X pattern requires one of the longest strings of moves used in this book.

$L^{-1}R^2U^2D^2B^2U^2D^2R^2L^2F^2L^{-1}$
Figure 43

$RL^2F^2B^2U^2R^2L^2F^2B^2D^2$
$RDU^2F^2B^2R^2U^2D^2F^2B^2L^2D$
Figure 44

$L^2R^2B^2L^2R^2B^2L^2F^2B^2R^2$
(or $(F^2R^2B^2L^2)^3$)

Figure 45

$(F^2R^2B^2)^2U^2D^2$
Figure 46

$F^2B^2U^2D^2R^2L^2$
Figure 47

$B^{-1}R^{-1}LU^{-1}DFB^{-1}U^2R^2L^2$
$F^2B^2 D^2RDU^2F^2B^2R^2U^2D^2$
$F^2B^2L^2D$
Figure 48

11. *Is there any other way of making a pattern with two plain faces and four X faces?*

12. *Find another pattern which has six X faces. (Try combining those that you know already.)*

6. Stripes

In Chapter 2 we showed how to create a pattern with four vertical stripes and two plain faces (Figure 7) and an even easier stripe pattern appears in Figure 13! Another extremely easy pattern is the combination of stripes and crosses shown in Figure 49. This comes from the string $F^2B^2L^2R^2$.

$F^2B^2L^2R^2$
Figure 49

A similar pattern with the other type of cross can be made with $(R^2F^2B^2)^2$ and this is shown in Figure 50.

A variation on the four vertical stripes is the combination of two vertical stripes and two horizontal stripes shown in Figure 51.

$(R^2F^2B^2)^2$
Figure 50

$F^2U^2B^2F^2U^2F^2$
Figure 51

(FBRL)^3
Figure 52

Instead of vertical and horizontal stripes we can have diagonal stripes. An example is shown in Figure 52.

13. *Can you make a pattern with six diagonal stripes like the one in Figure 53?*

If you apply the string U^2D^2 to the pattern of Figure 52 you will find the pattern of Figure 54.

Figure 53

(FBRL)$^3U^2D^2$
Figure 54

14. *The combination of spot and stripe faces shown in Figure 55 can be made with the string $U^2L^2R^2D^{-1}U^{-1}F^2B^2DU^{-1}$. A similar pattern is shown in Figure 56. This one can be made with a string of only four moves. Can you find it?*

15. *Can you find a way to make the pattern with six stripes shown in Figure 57?*

$U^2L^2R^2D^{-1}U^{-1}F^2B^2DU^{-1}$
Figure 55

Figure 56

Figure 57

29

7. Edge Moves

$R^2UFB^{-1}R^2F^{-1}BUR^2$

$R^2U^{-1}FB^{-1}R^2F^{-1}BU^{-1}R^2$
Figure 58

In Chapter 4 we gave a collection of moves which affect only corners. These moves can be a great help in getting the cube unscrambled quickly, but they are only half the story. You also need to know strings of moves which affect only edges. So in this Chapter we present some useful edge 3-cycles. Recall that a 3-cycle is a string of moves which 'cycles' three cubes and leaves the other cubes in place. We begin with eight strings of moves which cycle three edge cubes in the upper face. These are grouped in pairs just like the corner 3-cycles of Chapter 4: each string is the *reverse* of the other string of its pair. All of these moves affect the same three edge cubes. However, apart from moving the cubes, the strings shown in Figures 59, 60 and 61 also 'flip' a couple of edge cubes. This is indicated by the + signs in the diagrams.

R⁻¹F⁻¹R²D²L²B⁻¹L²D²R⁻¹   RD²L²BL²D²R²FR

Figure 59

FUF⁻¹L⁻¹B⁻¹R⁻¹U⁻¹RBL   L⁻¹B⁻¹R⁻¹URBLFU⁻¹F⁻¹

Figure 60

LFRU⁻¹R⁻¹F⁻¹L⁻¹B⁻¹UB   B⁻¹U⁻¹BLFRUR⁻¹F⁻¹L⁻¹

Figure 61

$F^{-1}BU^2FB^{-1}R^2$
Figure 62

$DR^{-1}LU^2RL^{-1}B^2D^{-1}$
Figure 63

One of the easiest 3-cycles to remember is shown in Figure 62. It is created by the string $F^{-1}BU^2FB^{-1}R^2$.

16. *There is another way of making the pattern in Figure 62 using exactly eight half turns. Can you find it?*

From a 3-cycle such as the one in Figure 62 lots of other 3-cycles can be obtained by forming *conjugates* as described in Chapter 4. This means that, if you you want to cycle any three edge cubes, you first bring them in any way you please to the positions affected by a 3-cycle that you already know. Let us use X to mean the string which puts the cubes into position and Y to mean the 3-cycle. As we mentioned in Chapter 4 X may disturb many other cubes. But this doesn't matter, because after you apply X you apply Y and then the *reverse* of X. The overall effect is that almost everything gets fixed up and only the cubes you were interested in get cycled around.

For example, you want to cycle the edge cube at the front of the upper face, the edge cube at the back of the upper face and the edge cube at the bottom of the right face. You could do this by taking X to be D and Y to be the string $R^{-1}LU^2RL^{-1}B^2$, which has the same effect as the string of Figure 62. Then XYX^{-1} is the string $DR^{-1}LU^2RL^{-1}B^2D^{-1}$. If you try this you should see the pattern shown in Figure 63.

There are many possible 3-cycles of edges and we would need over fifty more diagrams to present them all. Rather than do this we prefer to give a selection of a few of the shorter strings (Figures 64 to 67).

$(RU)^3(R^{-1}U^{-1})^2R^2$

$R^2(UR)^2(U^{-1}R^{-1})^3$

Figure 64

RU^{-1}R^{-1}F^{-1}L^{-1}B^{-1}U^{-1} RU^{2}R^{-1}F^{-1}L^{-1}B^{-1}
BLFRU^{2}R^{-1} UBLFRUR^{-1}

Figure 65

R^{-1}UR^{-1}LF^{2}RL^{-1}UR   R^{-1}U^{-1}LR^{-1}F^{2}L^{-1}RU^{-1}R

Figure 66

FR^{-1}D^{-1}UF^{2}DU^{-1}R^{-1}F^{-1} FRUD^{-1}F^{2}U^{-1}DRF^{-1}

Figure 67

You may have noticed by now that you can 'flip' two, four, six, eight, ten or twelve edges but never an odd number. So a cube with a single edge flipped would be an example of an impossible pattern.

The rather long string $F^2B^2LF^2D^{-1}UR^2BL^2R^2F^{-1}L^2DU^{-1}B^2R^{-1}$ flips four edges in a middle slice. Its effect is shown in Figure 68.

If this is combined with the string $(RLFBUD)^2$ which flips *eight* edges (Figure 69), then all *twelve* edges of the cube will be flipped (Figure 70).

The strings $R^{-1}U^2R^2UR^{-1}U^{-1}R^{-1}U^2LFRF^{-1}L^{-1}$ and $LF^{-1}UL^{-1}FB^{-1}UR^{-1}FU^{-1}RF^{-1}BU^{-1}$ flip exactly two edge cubes. From these it is possible to build up strings which flip any even number of edges anywhere on the cube.

$F^2B^2LF^2D^{-1}UR^2BL^2$
$R^2F^{-1}L^2DU^{-1}B^2R^{-1}$

Figure 68

$(RLFBUD)^2$

Figure 69

$F^2B^2LF^2D^{-1}UR^2BL^2R^2F^{-1}$
$L^2DU^{-1}B^2LFBUDRLFBUD$

Figure 70

8. Large and Small Twists

At the end of Chapter 4 we promised to give more information about *twisting* corner cubes and in this chapter we carry out that promise. Out of all the strings of moves which twist corners our favourite is $(R^{-1}D^2RB^{-1}U^2B)^2$. (Remember that the 2 means that the string inside the brackets is to be carried out *twice*.) This string, or something very like it, is said to have been discovered by Rubik himself and the pattern which it produces (Figure 71) can be seen on the front cover of the magazine **Scientific American** (March 1981). Its effect on the octagonal barrel is also rather striking.

If you have only looked at the illustration and not actually carried out the string you may be tempted to believe that only one corner has been twisted. In fact there is no string of moves whatsoever that can twist just one corner. So somewhere on the cube there must be another twisted corner cube. In the illustration, at least one square on seven of the eight corner cubes is visible, and only those on the corner cube at the upper right position of the front face are out of place. Consequently, the other cube which is twisted is the one at the bottom left position of the back face.

Instead of twisting just a pair of corners it is possible to twist a couple of 2 × 2 × 2 blocks, or at least to produce a pattern which has this appearance. A string which does this is $BL^{-1}D^2LDF^{-1}D^2FD^{-1}B^{-1}F^{-1}RU^2R^{-1}U^{-1}BU^2$ $B^{-1}UF$. The pattern is shown in Figure 72. This is the previous pattern on a larger scale!

$(R^{-1}D^2RB^{-1}U^2B)^2$
Figure 71

$BL^{-1}D^2LDF^{-1}D^2FD^{-1}B^{-1}$
$F^{-1}RU^2R^{-1}U^{-1}BU^2B^{-1}UF$
Figure 72

If you combine this string with the one of Figure 71 you will produce the pattern of Figure 73(i). Do it again and Figure 73(ii) will appear.

On the other hand, the 'six spot' string FB⁻¹UD⁻¹RL⁻¹FB⁻¹ applied to Figure 72 produces the pattern of Figure 74.

(i)

(ii)
Figure 73

Figure 74

$(BR^{-1}D^2RB^{-1}U^2)^2$
(i)

$R(RF^{-1}D^2FR^{-1}U^2)^2R^{-1}$
(ii)
Figure 75

In Figure 75 we show two ways of twisting a couple of corner cubes of the upper face. Both these strings were discovered by Rubik (and no doubt by many other people as well).

If you use the string of Figure 75 (ii), change the orientation of the whole cube and then use it again, it is possible to twist three corner cubes simultaneously. A shorter way of doing this is to use the move of Figure 76.

Using this string and the method of 'conjugates' desribed in Chapters 4 and 7 it is possible to twist three corner cubes anywhere on the cube.

$BRB^{-1}R^{-1}U^2L^{-1}B^2$
$LB^{-1}L^{-1}B^2LB^{-1}U^2$

Figure 76

9. Odds and Ends

Figure 77

Up until now our patterns have possessed quite a lot of symmetry. But a design doesn't have to be symmetrical to be interesting. In this chapter we present a few patterns which have nothing to do with spots, stripes, crosses, twists or flips. Most of these were discovered by Richard Walker. Instructions for making the first two appeared in the March 1981 issue of **Scientific American.** The pattern in Figure 77 is called the *worm* and the one in Figure 78 is called the *snake*.

The worm: $RUF^2D^{-1}RL^{-1}FB^{-1}D^{-1}F^{-1}R^{-1}\cdot$
$F^2RU^2FR^2F^{-1}R^{-1}U^{-1}F^{-1}U^2FR$

The snake:
$RFB^{-1}D^{-1}F^2DBF^{-1}R^{-1}F^2UR^2U^{-1}DF^2D^{-1}$

By changing the last part of the string which produces the snake we obtain a nice pattern of *circles* (Figure 79).

Figure 78

$RFB^{-1}D^{-1}F^2DBF^{-1}R^{-1}F^2U$
$R^2U^{-1}DR^2FBU^2F^{-1}B^{-1}$
R^2D^{-1}

Figure 79

DRF²R⁻¹F⁻¹D⁻¹R⁻¹D²RF²
DR²D⁻¹R⁻¹F⁻¹D⁻¹F²DR
Figure 80

R⁻¹D²FBR²F⁻¹D⁻¹RB⁻¹
DRL⁻¹F⁻¹DR⁻¹FD⁻¹R⁻¹
LD²F²R
Figure 81

L⁻¹R²F⁻¹L⁻¹B⁻¹UBLFRU⁻¹
R²FB⁻¹UD⁻¹RL⁻¹
Figure 82

In the next two patterns *large circles* of six edge cubes have been cycled around.

17. *There is a way of getting from the* **large circle** *pattern of Figure 80 to the* **worm** *using just eight moves. Can you find it?*

The remaining patterns have plain or U patterns on each face.

The string of moves given in Figure 83 has an interesting effect on the octagonal barrel. An even more interesting effect occurs when you use 'half' the string. Apply $RLU^2R^{-1}L^{-1}$ to the octagonal barrel and you will produce the pattern in Figure 84; try it and see.

18. *In Figure 84 two edge cubes appear to have been swapped. This is an impossible move for the cube. What is the explanation?*

$R^{-1}L^{-1}D^2RLF^{-1}B^{-1}D^2FB$
Figure 83

$RLU^2R^{-1}L^{-1}$
Figure 84

41

10. Doing the Impossible

It turns out that, no matter how hard you try, you will never succeed in finding a sequence of moves whose only effect is to swap a single pair of small cubes whilst leaving all the other cubes in their original spots. Many cube owners have tried to do this. They no doubt believed that the string they sought would be quite long, but that with perseverance and a little luck they would eventually find it. Unfortunately, they seek in vain, for the move is simply not possible. Others have tried to find a move which would flip a single edge cube. This too is impossible. And yet another impossible process is a twist of a single corner cube.

The fact that these things are impossible is not obvious. Indeed, it would take quite a while to give a complete account of the reasons why they cannot be done. This partly explains why some people keep searching for the 'lost moves'. But, for a few unlucky seekers, a more compelling reason to keep searching is that they have actually *seen* cubes displaying the 'impossible patterns'. How can this be? Fortunately, the answer is quite simple. The cube can be taken apart and then reassembled in any way you please, including the so called 'impossible' ways! Actually, some of the patterns which occur in our puzzles are still impossible, even if you try to make them by reassembling the cube. This is because they were made by peeling off the coloured squares and then rearranging them.

Taking the cube apart is quite easy. First rotate the upper face through 45° (Figure 85 (i)). Next use your thumb or a screw driver to prise out an edge cube from the upper face (Figure 85 (ii)). After doing this and twisting a few faces the cube will fall apart in your hands.

(i)

(ii)

Figure 85

42

To reassemble it, begin by putting back the edge pieces of the bottom layer (Figure 86 (ii)) and then the corner pieces of the bottom layer (Figure 86 (ii)). Next put back the edge pieces of the middle layer (Figure 87).

(i)

(ii)

(iii)
Figure 86

(i)

(ii)
Figure 87

(i)

(ii)

Figure 88

To put back the pieces in the upper layer you may need to turn the faces a little in order to get them in. (Figure 88 (i)). Remember to leave an edge piece to last and to have the upper face at 45 degrees to the other layers before you push the last piece in (Figure 88 (ii)).

Some cubes have designs (such as card suits) on the coloured squares (Figure 89). For these models it is important to have the six *centre* pieces (on the ends of the spindles) in their correct positions before you begin the reassembly process. (You should also be careful not to disturb their orientations while you are putting the pieces in.)

It is advisable to reassemble the cube in the plain pattern, otherwise you may find that it is impossible to get back to it just by rotations of the faces. In fact there are twelve (or twenty-four if your cube has designs on the coloured squares) essentially different ways of reassembling a cube, only *one* of which leads back to the original plain pattern.

Figure 89

11. The Cube in Disguise

There are now many puzzles on the market which are very similar to the original Rubik's Cube. Most of these puzzles are just cubes in disguise. That is, you can take them apart using the method described in the last chapter and inside you will find the mechanism of six spindles that is used to hold the cube together (Figure 86 (i)).

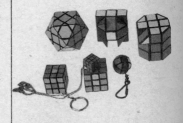

This is very convenient, because it means that it still makes sense to talk about the *up, down, front, back, left* and *right* faces and to use the same shorthand of U, D, F, B, L and R to describe the rotations of these faces. Some of the states (Figure 90) that these puzzles can get into can appear very strange, but after a while you get used to it!

Any method you know that unscrambles a cube can also be used to unscramble any of these puzzles. However, because the moving pieces are not always cubes, and because the colour schemes are different, it is not always obvious where some of the pieces should go. For example, some of the edge pieces of the octagonal barrel (Figure 91) have only one colour on them, but other edge pieces look just like the edge pieces on the cube. This means that if you flip an edge piece with a single colour and an edge piece with two colours it *appears* that the impossible has happened and only one edge has flipped. To do this, hold the barrel 'on its side' as in Figure 91 and apply the string RL⁻¹URL⁻¹BRL⁻¹DRL⁻¹F. If you want to see what this string *really* does, try it out on the cube.

Figure 90

RL⁻¹URL⁻¹BRL⁻¹DRL⁻¹F
Figure 91

Another move which is very useful when unscrambling the octagonal barrel is $RLU^2L^{-1}R^{-1}$. If the barrel is held upright, this appears to swap a single pair of edge cubes. Again you should try it on the cube to see what really happens.

Apart from changing the shape of the cube, manufacturers have put a variety of designs on the coloured squares that decorate their puzzles. This means that, in order to restore the original pattern on these models, you will need to know how to twist a few centre squares without disturbing the edge and corner pieces.

It is not possible to turn a *single* centre square through a *quarter* turn without disturbing other pieces, unless of course you take the cube apart and reassemble it. However it *is* possible to produce a single *half* turn of a centre square. To do this to the centre square of the *upper* face use the string $(URLU^2L^{-1}R^{-1})^2$.

This string is based on the one we used to swap two edge pieces of the octagonal barrel. Its effect on the cube is shown in Figure 92.

As well as knowing how to rotate a *single* centre square through a *half* turn, you also need to know how to rotate pairs of centre squares through *quarter* turns if you want to be able to get the cube back to its original pattern. Here is the complete solution.

$(URLU^2L^{-1}R^{-1})^2$
Figure 92

1. Find a centre square which is not in its correct orientation. Hold the cube so that this centre square is on the *upper* face as in Figure 92. If there are none to be found you have finished.
2. If the centre square on the upper face needs to be given a *half* turn, use the string $(URLU^2L^{-1}R^{-1})^2$ and then return to instruction 1.
3. If the square on the upper face needs to be given a *quarter* turn, and the square on the *down* face is not in its correct orientation, go to instruction 4. If the down face is correct and the centre square of the upper face needs a quarter turn, you will find that one of the centre squares of a side face also needs to be turned. Hold the cube so that this side face which needs to be fixed is the *right* face. If the

centre square of the upper face needs a *clockwise* quarter turn, use the string $R^{-1}UD^{-1}F^{-1}BRL^{-1}UR^{-1}LFB^{-1}U^{-1}D$; if it needs an *anticlockwise* quarter turn, use the string $UD^{-1}F^{-1}BRL^{-1}U^{-1}R^{-1}LFB^{-1}U^{-1}DR$. Now return to instruction 1.

4. If you have to use this instruction, it is because the centre square on the *upper* face needs a quarter turn and the centre square on the *down* face is also not correct. To give the centre square of the upper face a *clockwise* quarter turn, use the string $RL^{-1}F^2B^2RL^{-1}URL^{-1}F^2B^2RL^{-1}D^{-1}$. To give it an *anticlockwise* quarter turn use the string $RL^{-1}F^2B^2RL^{-1}U^{-1}RL^{-1}F^2B^2RL^{-1}D$. Return to instruction 1.

These instructions are guaranteed to restore the centre squares to their correct orientations, but they are certainly not the quickest way of doing it! To find short cuts, you need to know the precise effect of each instruction, and then choose the ones which best suit the situation. Here are a few examples.

a $(URLU^2L^{-1}R^{-1})^2$. This rotates the centre square on the *upper* face through a *half* turn.

b $R^2F^2B^2L^2D^2R^2F^2B^2L^2U^2$. This produces *half* turns of the centre squares of both the *upper* face and the *down* face.

c $RL^{-1}F^2B^2RL^{-1}URL^{-1}F^2B^2RL^{-1}D^{-1}$. This rotates the centre square of the *upper* face through a *clockwise quarter turn* and the centre square of the *down* face through an *anticlockwise quarter turn*.

d $R^{-1}UD^{-1}F^{-1}BRL^{-1}UR^{-1}LFB^{-1}U^{-1}D$. This rotates the centre square of the *upper* face through a *clockwise quarter turn* and the centre square of the *right* face through an *anticlockwise quarter turn* (Figure 93).

e $UD^{-1}F^{-1}BRL^{-1}UD^{-1}B^{-1}R^{-1}DR^{-1}LFB^{-1}U^{-1}$ $DR^{-1}BU$. This produces *half* turns of the centre squares of the *upper* and *right* faces.

f $(U^{-1}F^2R^2B^2D^{-1}B^2R^2F^2)^3$. This gives *clockwise quarter turns* to the centre squares of the *upper* face and the *down* face.

$R^{-1}UD^{-1}F^{-1}BRL^{-1}$
$UR^{-1}LFB^{-1}U^{-1}D$
Figure 93

12. Solutions to the Puzzles

1. (page 10) The string R^2L^2 changes the pattern of Figure 7 into the pattern of Figure 8.
2. (page 11) Apply the string $B^2D^2U^2F^2$ (which creates the pattern of Figure 9) to the pattern of Figure 10 and you will create a 'four spot' pattern.
3. (page 11) If you take the cube apart you will see that the centre pieces are attached to spindles which are in a fixed position. So it is not possible to swap two of these pieces and at the same time leave the others in place.
4. (page 14) The three patterns are: the plain pattern, the 'six spot' pattern and the 'four spot' pattern.
5. (page 16).This pattern is impossible to make by turning the faces but it *can* be constructed by dismantling the cube and reassembling it.
6. & 7. (page 16) The slice moves $UD^{-1}F^{-1}BRL^{-1}FB^{-1}$ create the pattern of Figure 22 and the slice moves $F^2B^2R^{-1}LU^2D^2RL^{-1}$ create the pattern of Figure 23.
8. (page 23) This cross pattern is also created by the slice moves $F^2B^2U^{-1}DF^2B^2UD^{-1}$.
9. (page 24) This is impossible, even if you try to do it by taking the cube apart.
10. (page 24) The required string of moves is $U(R^2F^2B^2)^2U^{-1}(R^2F^2B^2)^2$.
11. (page 26) In any 'four X' pattern the two plain faces must be opposite one another. But then the only possible pattern is the one shown in Figure 46. (If you take the cube apart and reassemble it you *can* make another type of 'four X' pattern.)
12. (page 26) If you first use the string of moves given in Figure 48 and then use $F^2B^2U^2D^2R^2L^2$ you will create a new 'six X' pattern. (There are others as well.)
13. (page 28) This can only be made by peeling off some colour patches and rearranging them.

14. (page 29) The string of half turns $D^2F^2B^2U^2$ produces the pattern you want.
15. (page 29) This pattern is not possible to make even if you take the cube apart and reassemble it. However, only three faces of a cube can be seen at once. So it is possible to create the *illusion* of this pattern by arranging just *three* faces correctly and leaving the other three faces (the ones you can't see) in disarray.
16. (page 32) The eight half turns $R^2F^2D^2B^2L^2B^2D^2F^2$ create the pattern of Figure 62.
17. (page 40) Begin with the cubes in the pattern of Figure 80, then apply the 'six spot' move $U^{-1}DFB^{-1}RL^{-1}U^{-1}D$ and you will find the 'worm'.
18. (page 41) Try out the move on the cube and you will find the pattern shown in Figure 94.

$RLU^2R^{-1}L^{-1}$
Figure 94

You can see that, apart from a pair of edge cubes being swapped, the whole of the 'front-right-column' has been swapped with the 'back-left-column'. On the octagonal barrel, it's hard to tell that this has happened, because the pieces of the barrel don't have 'home positions' the way they do on the cube.

PROBLEMS

Problems worthy
of attack
prove their worth
by hitting back.

Piet Hein